A Source Book of
Fire Engines

A Source Book of
Fire Engines

Denis Miller

Ward Lock Limited · London

© Ward Lock Limited 1983

First published in Great Britain in 1983
by Ward Lock Limited, 82 Gower Street,
London WC1E 6EQ, a Pentos Company.

House editor Suzanne Kendall
Layout by Sandra Buchanan

Text filmset in Univers

Printed and bound in Great Britain by
Netherwood Dalton and Co. Limited, Huddersfield

British Library Cataloguing in Publication Data

Miller, Denis
  A source book of fire engines.—(Source book
series)
  1. Fire-engines—Great Britain—History
  I. Title
628.9'25   TH9371

ISBN 0-7063-5895-3

Cover: Braidwood open-bodied Dennis 'Big Four' 1935, by
courtesy of Roger Pennington.

Frontispiece: This 'Braidwood'-bodied Shand-Mason horse-drawn
steamer could cope with a crew of ten. The basket arrangement at
the rear was used as a filter when taking water from pond or
stream.

Abbreviations

BA	breathing apparatus
bhp	brake horsepower
gpm	gallons per minute
GVW	gross vehicle weight
lpm	litres per minute
PE	pump escape
pto	power take-off
PWT	pump water tender
TL	turntable ladder

Introduction

Few aspects of everyday life can be quite as exciting as a fire brigade turnout and subsequent race to the incident which can be anything from a horse tangled in barbed wire to a major fire in a multi-storey hotel. Whatever the problem, the brigade must be capable of handling it and over the years specialist manufacturers have developed equipment to do just that.

The first recorded fire brigade was the Corps of Vigiles operating in Rome before the birth of Christ, but this was really an isolated case, for, with the fall of the Roman Empire, fire protection was virtually forgotten until the seventeenth century when some Continental inventors, such as the Dutchman Jan van der Heyden, developed the first manual pumps, attaching lengths of leather hose so that the pump itself did not have to be placed too close to the fire.

The Great Fire of London in 1666 did much to improve the situation, drawing the public's attention to the inherent danger of fire and prompting several insurance companies to organize their own brigades to deal with incidents involving their insured properties. Improvements were rapid at this stage, as each had to 'out-do' its rivals to secure new policy holders, and in 1721 a new type of pedestrian-hauled manual pump was patented by one Richard News-ham, a London button manufacturer, incorporating a built-in air chamber to equalize the pump's output and thus make it more efficient. Later, some insurance companies combined resources and gradually the old pedestrian-hauled appliances were superseded by horse-draught equipment, although still of the manual variety.

A long-wheelbase version of the 6×4 Morris-Commercial 'D'-Type chassis with 'Braidwood' body was used by Morris Commercial Cars Ltd to demonstrate this model's suitability as an off-road appliance.

While canals have long since ceased to be a viable answer to the transport problem, their proximity to industrial premises has often made them a convenient source of water at major incidents.

As late as 1968 Peter Pirsch, Sons & Co., Kenosha, Wisconsin, was still supplying massive bonneted aerial ladder trucks such as this artic with rear wheel steering.

The nineteenth century saw the introduction of the steam appliance, the first by Braithwaite & Ericsson, of Britain, in 1829. For a while, these were also horse-drawn but eventually steam propulsion appeared.

The world's first self-propelled steamer was constructed by Paul Rapsey Hodge, of New York, in 1840, bearing a close resemblance to a railway locomotive of the period, while the first British appliance of this type was the brainchild of William Roberts, of Brown, Lennox & Co., Millwall, London, whose cumbersome example was shown at the 1862 International Exhibition in Hyde Park. British manufacturers, however, were held back by the crippling 'Red Flag' Act which restricted road speeds by insisting that a man carrying a red flag walked ahead of any self-propelled vehicle to warn others of its approach, and it was not until the repeal of this act in 1896 that production models could be offered by Merryweather & Sons Ltd, of Greenwich, and

Sometimes even farm tractors were used to haul trailer-mounted pumps to fires in rural areas. This was one of the first British tractors, an Ivel, designed by Daniel Albone in 1897.

Shand, Mason & Co., of Blackfriars.

However, with the arrival of the internal combustion engine the steamer could no longer compete. The first successful petrol-driven motor pump was a trailer type demonstrated by German inventor Gottlieb Daimler at the Tunbridge Wells Exhibition & Demonstration of Motor Vehicles in 1895, while the first self-propelled pump was a Merryweather 'Hatfield' supplied to the French Rothschild Estate in 1904, followed within months by the first combination pump escapes with demountable wheeled escape ladders.

Some early pumps, particularly in the United States, combined the rear half of an old steam pumper and the front of a petrol-engined chassis, some manufacturers even supplying such conversion units from stock, while other brigades, notably in Germany, used battery-electric chassis kept permanently on charge for front-line availability, some even carrying an electric pump which could be plugged into a mains supply or run off the vehicle's batteries upon arrival at an incident.

Vehicle bodywork throughout this period was almost exclusively of the 'Braidwood' type in which the crew stood along the sides or sat facing outwards. This had been designed by James Braidwood, who had later become Chief Officer of the London Fire Engine Establishment. Although crewmen were frequently thrown from such appliances when travelling at speed, this layout remained the most common until the 1930s when first the 'New World' type, with crew facing each other inside an open body, and then the limousine appliance, made their debut.

For many years most European pumps had been mounted at the rear but in America were mainly mid-mounted, an arrangement that soon found favour in Britain as these new body styles and the increasing use of demountable wheeled escapes prevailed in the United Kingdom.

Particularly well suited to dealing with outbreaks of forest fire, this German-built Unimog U1300L 4×4 chassis has been fitted out as a lightweight PWT.

Built soon after the merger of French truck manufacturers Latil, Renault and Saviem in 1955, this 6×6 airfield fire crash tender was delivered to Orley Airport, Paris.

Even before these pumps had arrived, self-propelled hose cars and chemical engines had been introduced, the latter as rapid intervention 'first-aid' appliances to plug the gap between a steamer arriving and actually getting up sufficient steam to commence pumping. The chemical engine normally used the reaction between a bicarbonate of soda solution and sulphuric acid to form carbon dioxide gas which then ejected water from the vehicle's tank. The drawback of this was that, once started, the water could not be stopped until the chemical reaction had ceased! Later chemical appliances overcame this by using two tanks, one containing compressed air and the other water, projection of the latter being controlled by a tap between the two.

The familiar British wheeled escape ladder,

With pump mounted beneath the driver's seat, this was a triple combination appliance built about 1929 by the American Ward LaFrance Truck Corp.

invented by Abraham Wivell in 1836, was quickly adopted as standard by the Royal Society for the Protection of Life from Fire and, as we shall see in the main text of this book, manhandled to the site of the incident. The design was further improved in 1880 by James Shand, of Shand, Mason & Co., who added a pair of sheer-legs so that it could be converted into a free-standing water tower, a role it continued to fill until the adoption of the German-designed turntable ladder (TL) for this purpose. The first horse-drawn escape appeared in 1890 and the first motor-driven model some thirteen years later. As no power was necessary for pumping, the battery-electric escape carrier enjoyed brief popularity. Gradually, however, new pump escapes entered service or, where finance was not available, old escape vans or motor pumps were simply converted into pump escapes, the world's first new PE, a Merryweather, being supplied to London's Finchley Fire Brigade in 1904.

The American equivalent of the wheeled escape was the hook and ladder truck, while that of the turntable ladder was the aerial ladder. Both were available in rigid and articulated form, often with an

Delivered to the Scottish South Eastern Fire Brigade in 1963, this Merryweather emergency tender displays the many types of rescue equipment that must be carried.

Many early manual appliances were still giving good service at the turn of the century. This leather-hosed 'squirt' was used at Stoke Edith Park, Herefordshire, home of the Rt Hon. Lady Emily Foley.

extreme rear overhang in the case of the former or with a rear steersman's position to guide the rear wheels of the artic version round corners on city use. Most ladders were of timber construction although the first all-steel aerial ladder had been built as early as 1888 by the E. B. Preston Co., Chicago, Illinois. The first hydraulic-powered aerial ladder was constructed in 1936 by Peter Pirsch, Sons & Co., Kenosha, Wisconsin, and the world's first turntable ladder is said to have been a manually-elevated German-built Magirus mounted on a horse-drawn carriage in 1892. Manual TL elevation was replaced by the use of compressed carbon dioxide gas in 1901 and by full motor power five years later.

The increasing use of oil and its derivatives after World War I introduced even greater hazards than before, necessitating the design and construction of more specialist appliances to cope with emergencies involving such materials. The foam tender was just one of these, making its debut with the London Fire Brigade in 1925, and quickly adopted by others for use in an industrial environment.

The period leading up to World War II saw many new and revolutionary developments to combat the increasing risks of fire-fighting. Combination appliances, often incorporating the functions of three or more earlier machines, were becoming increasingly popular and the first airfield fire crash tenders, principally for military use, were introduced, some of the first being constructed in Germany about 1935. Indeed, it was Germany that took the lead in setting up a highly efficient modern brigade system with standardized equipment in 1935, the same year that the International Fire Prevention &

Using an ex-Services Reo-Studebaker 'M'-Series 6×6 chassis, fire authorities in Colorado, USA, constructed an appliance ideal for carrying water to forest fire outbreaks.

Public Security Exhibition was held in that country, enabling others to make careful note of such advances and gradually introduce them in many other parts of the world.

World War II did even more to revolutionize fire-fighting techniques. The effects of bomb damage in major European cities quickly made water supplies highly unreliable and frequently non-existent, and with no water the motor pump was useless. Thus, the mobile dam unit arrived, comprising a dropside lorry and demountable canvas or steel tanks which could be set up at the roadside and filled with water from any available source. Following this intermediate measure large numbers of water tenders and trailer pumps were introduced on standard commercial chassis and as time and facilities permitted so these tenders received their own pumps, often front-mounted, while the trailer pumps were used to maintain a good supply in the tender's tank.

The first production trailer pumps had appeared in the early 1920s, mainly for rural use where they could be coupled to any type of vehicle without tying

For many years French fire brigades favoured the 2-wheeled hand-hauled hose reel. Just like the British escape ladder, this was eventually mounted on quick-release securing points at the rear of motor appliances.

up expensive motor appliances. Again, wartime brought the trailer pump into its own and in Britain alone new trailer pumps outnumbered utility appliances almost 3 to 1! Some, such as the Beresford 'Stork', even carried a demountable pump so that the same 2-wheeled carriage could deliver further units to the fireground or when one pump was being serviced so another could be carried. By the war's end, however, the trailer pump had largely outserved its usefulness and from 1950 onwards an appliance not equipped with its own fitted pump was the exception rather than the rule.

The most popular replacement was the pump water tender (PWT), a development of the utility water tender and trailer pump combination. Again ideal for rural areas, where in some parts mains water was probably sadly lacking, the PWT has since developed into a highly versatile 'first-aid' appliance along with its derivative, the pump escape (PE). Early PWTs were either conversions of existing utility machines or, particularly on the European mainland, new machines based on rugged ex-Services chassis, many comprising only an austere tank body and rear- or mid-mounted pump.

A variation of both the PWT and the PE was the dual-purpose (DP) appliance, combining the attributes of both. This was ideal for small country towns that required both rural and urban machines but did not have the funds to employ separate appliances.

Post-war years have seen many other technological improvements, particularly those making appliances faster, more manoeuvrable and generally more efficient. While American fire departments had been training specialist units for many years, it was only now that some European brigades began looking at this and by the 1960s most of the world's brigades included specialist rescue squads for handling road accidents, rail crashes and similar incidents, using mainly high-mobility vehicles.

By the late 1960s many American aerial ladder trucks were being replaced by more useful and less complicated hydraulic platforms and in Britain similar units were beginning to oust the turntable ladder in some brigades. However, different Chief Officers continued to have different ideas on equipment and while some argued that the hydraulic platform could reach many windows that the TL could not, others insisted that the TL did not require as much space in which to operate. This controversy continues even today, although the hydraulic platform does now have the edge over the TL.

Built by the American Seagrave Corporation in 1910, this was a combination chemical engine and hose car with an underseat engine.

This '500'-Series Dodge K-1113, bodied and equipped as a PWT by Cheshire Fire Engineering Ltd, was one of a number delivered to the Royal Berkshire Fire Service in 1977/8.

Not generally recognized as a builder of fire-fighting vehicles, AB Volvo of Sweden, constructed a 6×6 version of its F89 range for operation as an airfield fire crash tender.

The increasing size of aircraft as air travel becomes more freely available has resulted in a complete overhaul of fire-fighting methods in this field. By the 1970s virtually all airfield brigades were introducing rapid intervention 'first-aid' units, usually based on comparatively light and very fast 4×4 chassis, copying the original chemical engine idea of earlier years. These machines are designed to commence rescue and fire-fighting work at least two or three minutes ahead of a totally new breed of massive fire crash tender with a phenomenal output of foam. While these heavier machines are frequently even faster than their 1960s counterparts, it is realised that in the case of aircraft fires just a few seconds count and if there is any way that fire-fighting equipment of any sort can reach an aircraft before a major crash tender then that is the best option.

The increasing cost of new specialist appliances has now led many brigades back to the days of the 1920s when many machines were mounted on standard commercial chassis. True, their modern counterparts have been considerably modified to bring them into line with Government requirements but in many cases the days of the specialist appliance chassis manufacturer are over except, perhaps, in the aircraft field. Similarly, the expense of replacing equipment, particularly ladders, means that more and more must reappear in conditioned form on new chassis and even turntable ladders and other specialist gear can now be re-used in this way.

What then of the future? There can be no doubt whatsoever that fire-fighting, particularly in industrial areas, will become increasingly scientific and any way of speeding the answering of fire calls will be jumped upon. Throughout the world fire alarm systems can now be linked to the nearest fire station but inevitably there are often more malfunctions than true fire calls. Computerization may well be the answer but on the fire appliance front itself the sheer density of city populations and clogging traffic make the fire-fighter's job an increasingly difficult one despite improvements in the tools of the trade.

Manual Pump
Hadley, Simpkin & Lott (GB) 1806

Messrs Hadley, Simpkin & Lott was the predecessor of one of Britain's most famous fire equipment manufacturers—Merryweather & Sons Ltd, Greenwich, London. Typical of the earlier company's models was this manual pump to King's patent but based on Richard Newsham's design. The business had been founded by Samuel Hadley when he acquired Adam Nuttall's fire engine factory in London's Long Acre in 1769, being joined by engineer Charles Simpkin in 1792 and later re-named Hadley, Simpkin & Lott when Henry Lott, the son of Squire Lott, of Twyford, Berkshire, became senior partner upon the death of Simpkin and Lott's marriage to his widow. The firm's manual pumps were quite an advance over earlier types, having a steerable fore-carriage for manual- or horse-draught, as well as the usual handrails for pumping on each side of the appliance. There was provision for a second set of volunteers who stood on the machine, holding on to the upper fixed handrails, so they could step on and off an additional set of treadles in time with the main pumping action in order to boost the machine's power. Water in the lower trough was maintained either by bucket chain or by leather suction hose.

Horse-drawn Steam Pump
Merryweather (GB)   1863

The second National Steam Fire Engine Contest was held at the Crystal Palace, South London, in July 1863 where Merryweather & Sons Ltd (formerly Messrs Hadley, Simpkin & Lott) entered two horse-drawn appliances. Smallest was the *Torrent*, a single-cylinder model, while the largest was a vertical-boilered double horizontal-cylinder design called the *Sutherland*. This machine proved most successful, projecting a good 49/52 m (160/170 ft) jet of water through a 38 mm (1½ in) nozzle and receiving 1st Prize for its efforts. The Admiralty was particularly impressed, purchasing the engine immediately after the contest for use in the Naval Dockyard at Devonport, near Plymouth, where it was on active service until 1905. For a few years it stood in the Merryweather museum back at Greenwich but was presented to the Science Museum in 1924 and is now the world's oldest surviving steam fire appliance.

Horse-drawn 'Curricle' Escape
Shand-Mason (GB) 1874

The other leading British manufacturer of this period was Shand, Mason & Co. Ltd, of Blackfriars, London, which had been founded in 1851 although its origins went back as far as 1774. This 12 m (40 ft) 'Curricle' escape ladder was used in and around the Cliveden Estate in Berkshire for many years. It was equipped for horse-draught, was hand-operated and the ladder could be removed from the carriage if required. For town use, such ladders were often pedestrian-hauled, usually with the ladder in a horizontal plane but sometimes 'balanced' at an angle! By the turn of the century many large cities had established street escape stations where one of these ladders was positioned and any member of the public who assisted a police officer in rushing the escape to the scene of a fire was handsomely rewarded. The escape shown is part of the Lord Leigh Collection, currently on loan to the museum of the West Midlands Fire Service as is the appliance in the background, a horse-drawn 8-man manual pump, also by Shand-Mason, constructed in 1851.

Steam Pumper
Amoskeag (USA) 1894

Supplied to the City of Hartford, Connecticut, this self-propelled Amoskeag was one of 22 similar machines built by the Amoskeag Manufacturing Co., Manchester, New Hampshire, between 1867 and 1908. The design called for a seamless vertical copper-tube boiler, 2-cylinder horizontal engine with cylinders of 241 mm bore × 203 mm stroke (9½ × 8 in), double chain-drive to rear axle and a 6,592 lpm (1,450 gpm) pump. It was capable of propelling a single jet some 107 m (350 ft) and in operational condition tipped the scales at over 7,257 kg (7 tons 3-cwt). To ensure a rapid getaway on receipt of an alarm call, the Hartford machine was kept permanently connected to a boiler in the engine house basement, thereby ensuring that the water was always hot, thus requiring only a couple of minutes to generate sufficient steam to go on the run. Said to be capable of speeds up to 50 kmph (31 mph), this appliance was usually limited to only 19 kmph (12 mph) because of the driver's inability to steer the lumbering monster at higher speeds and the effects of its vibration upon surrounding buildings!

Horse-drawn Steam Pump
Shand-Mason (GB) 1894

Maintained in full working order by the Berkshire Veteran Fire Engine Society, formed by employees of the Royal Berkshire Fire Brigade, this vertical-boilered Shand-Mason was supplied to the Englefield Estate, near Reading, in 1894. Described in the original order as 'an improved patent double-cylinder vertical steam fire engine of No. 1 size, capable of discharging 1,591 lpm (350 gpm) and throwing a jet to 49 m (160 ft) height through a 28 mm (1⅛ in) jet pipe', it was withdrawn from active service in 1940, lying unused until 1957 when it was transferred on a permanent loan basis to the then Berkshire & Reading Fire Brigade. Now fully restored, it holds a current boiler certificate and is used for demonstrations throughout the rally season.

Trailer Pump
Daimler (D)   1895

The first motorized pump in Britain was delivered to the Hon. Evelyn Ellis for use on his private estate at Datchet, Buckinghamshire, in 1895. It was not the world's first, however, the same German firm building this, for manual propulsion, in 1888 and another, delivered to Cannstatt, Germany, in 1892. The British machine demonstrated its skills in the hands of the local volunteer fire brigade at the Tunbridge Wells Exhibition & Demonstration of Motor Vehicles on 15 October 1895, proving highly efficient despite the misgivings of steam supporters. It was a horizontal single-throw belt-driven appliance intended for horse or pedestrian haulage but may well have been used in conjunction with the owner's Daimler car, the first passenger vehicle to be imported into the British Isles, also in 1895. In this picture, the Hon. Evelyn Ellis is the bearded gentleman to the immediate right of the pump.

Horse-drawn 'Bangor' Hook & Ladder Truck
Seagrave (USA)   1900

In Britain the wheeled escape was the main piece of rescue equipment whereas American brigades relied at this time upon the hook and ladder truck. The Seagrave Co., Columbus, Ohio, offered a variety of types of which the 'Bangor' was representative. This particular example was delivered to Pond Creek, Oklahoma, in 1900, carrying scaling ladders for heights of up to 13·7 m (45 ft). As time went by, the length of such appliances increased dramatically, even to as much as 26 m (85 ft), whereupon they were re-designated aerial ladders, fitted with a rear steering position and spring-actuated raising and luffing motions to save time and energy. The appliance shown is owned by the Oklahoma Firefighters' Museum in Oklahoma City.

Motor Tender
Bijou (GB)   1901

In April 1901 the Eccles Borough Fire Brigade in Manchester placed Britain's first official order for a petrol-engined self-propelled appliance. They took their custom to the local Protector Lamp & Lighting Co. Ltd, builder of Bijou light passenger cars since the previous year. At a cost of £180, a motor tender was handed over on Monday 8 September 1901, having a 7 hp horizontal 2-cylinder engine, 2-speed transmission and chain final-drive. While maximum speed was claimed to be 23 kmph (14 mph), this appliance proved to be severely underpowered when carrying its full complement of five crew, ladders, hoses, standpipes and extinguishers, and it was short-lived.

Chemical Escape
Merryweather (GB)   1903

The world's first motor-driven escape was designed by Supt S. M. Eddington of the Tottenham Fire Brigade and assembled by Merryweather & Sons Ltd in 1903, this being the first petrol-engined vehicle constructed by the company. Powered by a 24/30 hp 4-cylinder engine located beneath the driver's footboard, it employed numerous Aster components imported from Ateliers de Construction Mécanique l'Aster, St Denis, France, and as well as a 15 m (50 ft) demountable wheeled escape ladder carried a 273 litre (60 gallon) chemical extinguisher and fixed hose reel. It was based at the Harringay Fire Station, the first in Britain to be constructed specifically for motorized appliances. It was delivered in the form shown but the engine overheated badly and a coil-type radiator was added later.

Chemical Pump Escape
Merryweather (GB) 1904

Following Tottenham's lead, the nearby Finchley Brigade, under Chief Officer Sly, took delivery of the world's first combined motor pump escape and chemical engine, also a Merryweather, in November 1904 at a cost of nearly £1,000. Once again Aster components were used but the 30 hp petrol engine proved to be underpowered and was quickly replaced by a more powerful unit, necessitating some re-jigging at the forward end. Fire-fighting equipment comprised an 18 m (60 ft) demountable wheeled escape, 273 litre (60 gallon) chemical extinguisher with 55 m (180 ft) of hose and a shaft-drive triple-barrel 2,273 lpm (500 gpm) Merryweather 'Hatfield' pump mounted just forward of the chain-driven rear wheels. Withdrawn from service in 1928, this appliance was passed to the Science Museum, restored as shown, and added to the museum's static collection.

Chemical Engine & Hose Car
Packard (USA) 1907

As a 'first-aid' appliance, the combination chemical engine and hose car was a valuable asset for any brigade, providing an immediate supply of water at high pressure. In 1907 the City of Ventnor, New Jersey, took delivery of just such a machine based upon the first commercial model, a 1,524 kg (1½ ton) job, built by the Packard Motor Car Co., Detroit, Michigan. Chemical equipment comprised two 273 litre (60 gallon) copper cylinders mounted trans-versely behind the driver, one containing water and the other compressed air which, when a tap was opened, expelled sufficient water at whatever pressure might be necessary. The 15 hp horizontal twin-cylinder petrol engine, 3-speed sliding-mesh transmission, double chain-drive and touring car back axle provided a maximum speed of 40 kmph (25 mph). Bodywork was by James Boyd & Bros Inc., Philadelphia, Pennsylvania, and suspension was by three semi-elliptic leaf springs, two mounted longitudinally at the front and the third transversely at the rear.

Model AC-30 Chemical Engine & Hose Car
Seagrave (USA)   1907

Seagrave's first motorized appliance, a combination chemical engine and hose car, was constructed in 1907. The company had worked on this design for over two years, assisted by vehicle manufacturers the Frayer-Miller Co. and the Oscar Lear Automobile Co., both also of Columbus, Ohio. Designated the Model AC-30, the prototype shown here had a 24 hp 4-cylinder vertical petrol engine cooled by a blower driven off the engine itself and thus operating independently of the vehicle's road speed. Mounted beneath the driver's seat, the engine was readily accessible via doors on each side of the vehicle and the 4-speed selective transmission drove the rear wheels via double roller chains. In this instance, a crank at the forward end of the hosebox tilted a chemical container to initiate a reaction between sulphuric acid and a bicarbonate solution, generating sufficient carbon dioxide gas to expel water under pressure. On 27 June 1907 this appliance undertook an 88·5 km (55 mile) test run from Columbus to Chillicothe, Ohio, and was subsequently shown at various fire chief's conferences before delivery to a Canadian brigade.

Model A.3 Chemical Engine & Hose Car
Albion (GB)   1910

The 12 hp vertical twin-cylindered Albion A.3 had been constructed since 1903 in the new South Street, Scotstoun, Glasgow, factory of the Albion Motor Car Co. Ltd. This combination chemical engine and hose car was constructed on an up-rated (16 hp) version of the 762 kg (¾ ton) capacity commercial chassis in 1910 and shipped to Australia for the Milton Volunteer Fire Brigade, of Brisbane. Here, the chemical cylinders were mounted longitudinally and instead of the more familiar warning bell for the home market the appliance carried a gong beside the driver. Transmission was via a 3-speed gearbox, bevel-drive to a differential countershaft and then chain final-drive to the rear wheels.

Motor Pump
Halley (GB) 1910

For less than three years another Scottish manufacturer, Halley's Industrial Motors Ltd, of Yoker, Glasgow, offered various types of fire appliance from a light 2-cylinder model up to a 65/70 hp 6-cylinder machine. The Leith Fire Brigade took delivery of one of the latter in 1910, this also having an Heli Shaw type clutch, 4-speed gearbox and double roller chain final-drive. A 2,046 lpm (450 gpm) Mather & Platt centrifugal turbine pump was located at the rear but when pumping the radiator capacity was found to be insufficient to cool the engine adequately so additional water was 'bled' from the main pump, diverted to the radiator and disposed of via an overflow. Bodywork was of the popular 'Braidwood' type with single extension ladder carried aloft. This appliance is now preserved by the Lothian Fire Brigade.

Motor Pump
Leyland (GB)   1910

Established in 1907, Leyland Motors Ltd, of Leyland, near Preston, Lancashire, built its first fire appliance in 1910 to the special order of the Dublin Fire Department. This was a high-speed motor pump powered by an 85 hp 6-cylinder petrol engine capable of producing a top speed of 97 kmph (60 mph). Like the Australian A.3 Albion, this carried a warning gong rather than a bell. Shaft-drive was employed, with a special gearbox power take-off driving the heavy-duty rear-mounted pump. All wheels were shod with solid rubber block tyres, unusual at this time, and the appliance was still in use as late as 1940, albeit on pneumatic tyres.

Type 5 Hose Car
American LaFrance (USA) 1911

The American LaFrance Fire Engine Co. was founded at Elmira, New York State, in 1903, combining the names of the LaFrance Fire Engine Co., of Elmira, and the American Fire Engine Co., of Seneca Falls. Initially, only steam appliances were built but in 1910 the company's first petrol-engined combination was demonstrated and by the following year hose cars such as this Type 5 were among the many types listed. Apparently powered by a specially constructed 4-cylinder Simplex engine, chain-drive appliances of this type were extremely fast, carrying ready-coupled lengths of hose in special compartments beneath the longitudinal bench-type crew seating. Formerly with the Sandy Hollow Fire Department, this machine is now in the Long Island Automotive Museum, Southampton, New York State.

High-Pressure Pumper & Hose Car
Gramm (USA) 1911

Anxious to improve its fire cover by increasing available water pressure, the New York Fire Department took delivery of an open high-pressure pumper in 1911. It was based on a chain-drive 3,048 kg (3 ton) 'cabover' chassis constructed by the Gramm Motor Truck Co., Lima, Ohio, and was bodied and equipped by the Webb Motor Fire Apparatus Co., St Louis, Missouri, which had been founded three years earlier specifically to construct self-propelled appliances. Hydrant inlets were located behind the nearside front wheel, with a rotating turret pipe mounted above. The narrow body contained flat lengths of ready-coupled hose and nozzles of varying sizes were carried on racks at the rear. For travelling, the crew stood somewhat precariously on top of the body, holding on to the metal framing provided. Engine access was via side panels and an unusual hinged radiator. The 'cyclops' headlamp was a distinctive feature, popular throughout America at the time.

Chemical Engine & Hose Car
Harder (USA) 1911

Powered by a 4-cylinder Waukesha petrol engine driving the rear wheels via a 3-speed transmission and double roller chains, this Harder 'cabover' combination chemical engine and hose car, with general equipment cage behind the driver, was delivered to the Chicago Fire Department by the Harder Fire Proof Storage & Van Co., also of Chicago, Illinois, in 1911. Elaborately finished, it was probably one of the earliest Waukesha-powered fire trucks and was certainly one of only a handful of appliances built by the Harder company which had originally constructed vehicles solely for its own use.

'YC'-Type Motor Pump
Commer Car (GB)   1913

Also known as the Commer-Simonis, this motor
pump appliance was equipped and bodied by Henry
Simonis & Co. to the order of the London Fire
Brigade using a 'YC'- or 'Barnet'-Type chassis constructed by Commercial Cars Ltd, Luton, Bedfordshire. Power came from a 24 hp 4-cylinder petrol
engine of own make, driving the rear wheels via a
leather-faced cone clutch, 3-speed gearbox, differential countershaft and side chain-drive running in a
patented oil-tight casing. Based at the Shooters Hill
Fire Station, this particular appliance remained with
the brigade until 1924 when it was returned to the
manufacturer, re-conditioned and sold to Chivers &
Sons Ltd, Cambridge, as a works engine. In 1961 it
was presented back to Commer Cars Ltd, the original manufacturer's successor, fully restored and
displayed in the Rootes Collection. Eventually it was
sold to enthusiast Ken Senior, of Addlestone, Surrey.

Steam Pump
Gobron-Brillié (F)   1913

This unique appliance was constructed for use on a country estate in 1913, using a 1907 Gobron-Brillié passenger car built by Soc. Gobron-Brillié, Boulogne-sur-Seine, and a vertical-boilered Merryweather steam pump. Power was provided by a most unusual 4-cylinder side-valve opposed-piston engine of 7·6 litres (464 cu in) capacity with two pistons per cylinder in which the explosions took place between them. It was claimed that such an engine could run on any fuel, including brandy or whisky! Transmission comprised a 4-speed gate-change box and chain final-drive, with a footbrake acting upon the transmission line and a handbrake upon the rear wheels. Of uncertain vintage, the vertical-boilered steam pump was located over the rear axle. This appliance is now in the ownership of the National Motor Museum, Beaulieu, Hampshire.

'N'-Type Motor Pump
Dennis (GB) 1914

New to the Greenall Whitley Brewery, Warrington, Cheshire, in 1914, this 'N'-Type appliance built by former bicycle manufacturers Dennis Bros Ltd, Guildford, Surrey, and now preserved by the Dennis Apprentices' Association, had a 50 hp 6·6 litre (403 cu in) 4-cylinder White & Poppe petrol engine, cone clutch, 4-speed 'crash' gearbox and Dennis patented overhead-worm rear axle. It was based on a standard 4,572 kg (4½ ton) commercial chassis with footbrake acting upon the transmission line and handbrake upon the rear wheels. A 1,818 lpm (400 gpm) Gwynne turbine pump was mounted at the rear. Used by the brewery until 1950, it was acquired by local Dennis distributor The Old Trafford Motor Engineering Co. Ltd, of Manchester, who presented it back to the manufacturer in 1960. It still has its original wooden wheels, although the solid rubber tyres have been renewed.

Chemical Engine & Hose Car
White (USA) 1916

Now preserved in Britain by enthusiast Ken Senior, of Addlestone, Surrey, who purchased it from the Long Island Motor Museum in 1980, this combination chemical engine and hose car was built by the White Motor Co., Cleveland, Ohio, in 1916 for the Bronx Fire Department, New York City. Considerably modified over the years, including the fitting of larger-section pneumatic tyres, this appliance remained on active service until 1947 when it passed to the museum. This White chassis was particularly popular with fire engine builders, with its distinctive radiator design inherited from the days of the White steam car.

Hose Car
Kelly-Springfield (USA) 1918

Supporting local industry, the Springfield Fire Department, Ohio, purchased this hose car based on a 'coal-scuttle' bonneted 1,016 kg (1 ton) Kelly-Springfield chassis built by the local Kelly-Springfield Motor Truck Co. in 1918, the same year that the Department also took delivery of a combination chemical engine and hose car on an identical chassis. In both cases the radiator was positioned between engine and driver, other features including a 3-speed transmission and double chain-drive, by now fairly unusual for an appliance of this relatively low capacity.

Chemical Engine & Hose Car
Dependable (USA) 1921

The Dependable chassis, built between 1914 and 1925, was exceptionally rare in truck form but even more so as a fire appliance. Based on a 1,524 kg (1½ ton) model, this combination chemical engine and hose car was assembled in the Dependable Truck & Tractor Co.'s Galesbury, East St Louis, Illinois, factory in 1921, using a Buda petrol engine, 3-speed Fuller transmission and worm-drive Wisconsin rear axle. Delivered to the Hartland Fire Department, it was equipped by the Boyer Fire Apparatus Co., Logansport, Indiana, following traditional American lines.

'M'-Type Pumper
Ahrens-Fox (USA)   1927

Regarded as 'the Rolls-Royce of the American fire service', the Ahrens-Fox was assembled by the Ahrens-Fox Fire Engine Co., Cincinnati, Ohio, the most famous design incorporating a heavy-duty pump located ahead of the engine, its most distinctive feature being a large spherical brass air chamber mounted on top. It was said that this was beaten to shape by hand and as only one employee was capable of doing this his method died with him and later appliances had to be fitted with a 2-piece forged steel ball. Up to 1927 the company's own monobloc 6-cylinder petrol engine was used, the 4-cylinder double-acting pump having a capacity of 3,409 lpm (750 gpm). On this particular appliance one branch of the front 'Y' intake was coupled permanently to a length of soft-suction hose stowed along the far running board to speed connection to a nearby water supply.

Model 'AA' Hose & Ladder Truck
Ford (USA)   1928

Delivered to the Soestdijk Fire Brigade, Holland, in 1928, this stretched 1,016/1,524 kg (1/1½ ton) Model 'AA' Ford acted as a 12-man hose and ladder truck. It carried Continental-style demountable rear hose reels. Built by the Ford Motor Co. at its Dearborn, Michigan, plant, this appliance used a 4-cylinder 3·3 litre (201 cu in) coil-ignition petrol engine, a short-lived worm final-drive and single rear tyres that were standard equipment at that time. The company's traditional transverse leaf spring suspension was employed. By the following year a 4-speed transmission, spiral-bevel drive and dual rear tyre option were listed, longitudinal cantilever-type rear suspension units being offered by 1930. The mass availability of the 'AA' at a considerably lower price than premium chassis in the United States led to some popularity amongst the specialist fire appliance builders who used the 'AA' for pump outputs of up to 3,409 lpm (750 gpm).

Motor Pump & Hose Car
Austro-Fiat (A)   1928

Founded in 1907, Österreichische Fiat Werke A.G. was established in Vienna to assemble Fiat cars and commercial vehicles for the Austrian and Hungarian markets. In 1925 this arrangement was terminated and in 1928, by then re-organized as Österreichische Automobil-Fabriks A.G., this 50 hp combination motor pump and hose car was one of a number of appliances constructed to the special order of various Austrian brigades. Fitted out by the well known European fire-fighting equipment specialist Rosenbauer, this carried a 4-stage pump with a capacity of up to 1,000 lpm (220 gpm) in front of the engine, two racks of rolled hose and a set of ladders.

Salvage Tender
Chevrolet (GB)   1929

Chevrolet trucks were built in Britain between 1928 and 1931 when they were phased out in favour of the Chevrolet-inspired Bedford manufactured at Vauxhall Motors Ltd, Luton, Bedfordshire. One Hendon-built Chevrolet chassis/cab found its way to the Fire Salvage Association of Liverpool in 1929, receiving a 1911-vintage 'Braidwood' body removed from an earlier Dennis. Full-scale fire-fighting equipment was not, of course, carried, the vehicle's activities being largely confined to clearing water and fire damage and generally securing fire-damaged premises. For clearing water a portable Dennis pump was carried while other equipment was by John Morris.

'G'-Type Motor Pump
Dennis (GB) 1929

Introduced as a low-loading passenger chassis with underslung worm rear axle in 1927, the 3·6 m (142 in) wheelbase Dennis 'G'-Type first appeared as a fire appliance a year later. This 'Braidwood'-bodied motor pump went to the Wotton-under-Edge, Gloucestershire, Fire Brigade in 1929, having a 4-cylinder Dennis petrol engine producing a nominal 17·92 hp, a cone clutch, 4-speed lightweight 'crash' box and 4-wheel mechanical servo-assisted brakes. Now preserved by the manufacturer's Apprentices' Association, this machine has a 1,136 lpm (250 gpm) Dennis turbine pump at the rear.

Emergency Tender
Dennis (GB) 1929

Based on the Dennis 1,524 kg (1½ ton) commercial chassis, both the Brighton and London Fire Brigades took delivery of specially constructed box-vans for use as emergency tenders, the former in 1928 and the latter in March 1929. Both carried a dynamo and cab bulkhead switchboard plus three emergency floodlights as shown, other equipment on the London machine including 'Proto' breathing apparatus, a 'Novit' oxygen revival set, oxy-acetylene cutting gear, hydraulic jacks and a comprehensive first-aid box. This model had a wheelbase of 3·4 m (132 in) and was powered by a 4-cylinder Dennis side-valve petrol engine of 17·9 hp output. Transmission comprised a cone clutch, 4-speed gearbox and overhead-worm rear axle.

Hose Car
Bedford (GB)   1932

By 1932 production of a bonneted 2,032 kg (2 ton) truck chassis was well under way in the Luton plant of Vauxhall Motors Ltd. As we have already seen, this was based on Chevrolet design, still incorporating 6-volt coil ignition, 4-wheel mechanical brakes and spiral-bevel drive but Anglicized by the use of a new 3·2 litre (195 cu in) 6-cylinder petrol engine producing some 44 bhp and a 4-speed transmission. One of these chassis, minus cab, was used as the basis for a light hose car supplied to the Grayshot & Hindhead Volunteer Fire Brigade. Because no pump was fitted, this appliance usually hauled a Dennis trailer unit.

D4 Hose Car
Perl (A)   1932

Another rarity on the fire appliance scene was the Austrian-built Perl, constructed by Automobilfabrik Perl A.G., of Leising, Vienna. This D4 hose car, with extra crew seating and overhead extension ladder, was new in 1932, having a 50 hp 4-cylinder diesel engine. It was actually based on a low-loading passenger chassis providing easy access to all areas of the body. Note the location of the rear storage rack for rolled hoses. It was placed at this height so that each crewman could carry a roll on his shoulder to whatever location it was required—a much easier task than carrying it any other way. This machine is now preserved in a fire service museum.

'BC'-Type Escape Van
Thornycroft (GB) 1932

By the early 1930s there were two distinct types of commercial chassis available from British manufacturers—the low-loading passenger type and the straight-frame goods model—and increasingly the former was being adapted for more specialist tasks. One such example was this Thornycroft 'BC'-Type, built at Basingstoke, Hampshire, by John I. Thornycroft & Co. Ltd and supplied as a heavy-duty escape van to the Borough of Congleton, Cheshire. Bodywork was by the Lawton Motor Body Building Co., Stoke-on-Trent, of 'New World' style, with twelve crewmen sitting opposite each other on longitudinal bench seating. The driver sat in a half-cab along typical passenger vehicle lines and an 18 m (60 ft) wheeled escape and 1,364 lpm (300 gpm) Dennis trailer pump were both loaded on to the vehicle from the rear.

F7B Pump Escape
Scammel (GB)   1933

The early 1930s saw the entire Western World suffering from an acute unemployment problem following the 'slump' and Watford, was no exception. It was even feared that the famous plant of Scammell Lorries Ltd, one of the town's biggest employers, would have to close through lack of orders so, despite the company's non-existant experience in the fire-fighting field, Mr D. Spence, then Chief Officer of the Watford Brigade, recommended that the next appliance order should be placed with the firm and in September 1932 his successor, Mr S. B. Manning, did just that. This was an F7B pump escape powered by an 85 bhp 4-cylinder overhead-valve Scammell petrol engine specially modified to take an electric heating device in the water circulation system which could be plugged into a mains supply to ensure rapid starting in emergency. Fire equipment included a 2-stage 1,818 lpm (400 gpm) turbine pump by J. Stone & Co. Ltd, 36·5 m (120 ft) hose reel running off a 182-litre (40-gallon) copper 'first-aid' tank and a 16·8 m (55 ft) Bayley's demountable wheeled escape ladder. On delivery day, 3 February 1933, the appliance was put through its paces in front of a distinguished audience of civic

and other dignitaries at the Watford Waterworks but ended its days ignominiously in a local scrapyard in the 1950s.

'Ace' Motor Pump
Dennis (GB) 1934

The first Dennis appliances of 'New World' style appeared in 1933, based on a modified 2·9 m (9½ ft) wheelbase version of the bonneted 'Ace' commercial chassis, powered by a 4-cylinder Dennis petrol engine with a nominal output of 24·8 hp. Other mechanical details included a single-plate clutch, 4-speed gearbox, spiral-bevel rear axle and hydraulic 4-wheel brakes. This open motor pump was supplied to Oldbury, Warley, West Midlands, in 1934. A 1,590/2,045 lpm (350/450 gpm) No. 2 Dennis pump was mounted at the rear and in this particular instance a short scaling ladder was carried on the offside. Machines such as this were more common in rural areas, often in conjunction with a trailer pump, where their short wheelbase made them highly manoeuvrable in narrow lanes or congested farmyards.

D3 Pump Escape
Dennis (GB)   1934

Holder of the fastest road vehicle for the London-Paris Run in 1967, this Dennis D3 pump escape with old 'Braidwood'-style body was new to the Letchworth, Hertfordshire, Fire Brigade in 1934. Based on a low-loading bonneted chassis with a wheelbase of 3·8 m (12½ ft), it had a 5·7 litre (348 cu in) 4-cylinder side-valve Dennis petrol engine, 2-plate dry clutch, 4-speed gearbox and worm-drive rear axle, while brakes were of the 4-wheel servo-assisted type. A 1,590/2,045 lpm (350/450 gpm) 2-stage Dennis turbine pump was fitted at the rear.

Turntable Ladder
Laffly (F)   1934

By the early 1930s commercial chassis built under the Laffly name by Ets Laffly, Asniéres, Seine, were beginning to make inroads into the French fire appliance market in both diesel and petrol form. This 4-stage 30 m (98 ft) German-built Metz turntable ladder was mounted on just such a chassis, believed powered by the company's own 40 bhp petrol engine, for the Ville de Mortluçon brigade. Devoid of any other equipment, a 5-man crew could be carried, two in the open cab and the other three with their backs to them and facing the rear. A turret nozzle was mounted at the head of the ladder but when fully extended extra pumping equipment had to be brought in to maintain sufficient pressure.

'Terrier' Pump Escape
Leyland (GB)   1934

Definitely out of the ordinary for the Middlesex Fire Brigade, this bonneted Leyland 'Terrier', originally developed in 1928 as a 3,048 kg (3 ton) military model, was delivered in 1934. It featured a mid-mounted pump, 'Braidwood' body and demountable wheeled escape ladder and, as seen here, even carried two sets of breathing apparatus (BA) equipment. A double-drive bogie was fitted and power was believed to have come from a 6-cylinder Leyland petrol engine. While most 6-wheeled appliances were built at that time for overseas use, this machine was chosen because of its ability to carry the maximum amount of equipment within its increased overall length when compared with a 4-wheeled appliance.

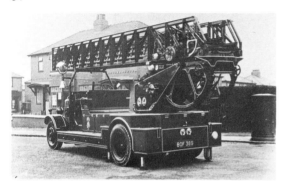

'TLM' Turntable Ladder
Leyland (GB)　1935

Despite an increased requirement for new efficient fire-fighting equipment in the years leading up to World War II, production of the Leyland 'TLM' turntable ladder appliance was curtailed somewhat prematurely early in 1940 for the simple reason that the turntable equipment was German! Introduced in late 1932, the standard specification called for a wheelbase of 4·4 m (14½ ft), a 115 bhp 6-cylinder overhead-valve petrol engine or, from 1935, an 8·6 litre (525 cu in) Leyland diesel unit. Transmission

comprised a single-plate clutch, 4-speed 'crash' box and underslung-worm rear axle. Ladder equipment was by Metz, normally of 26, 27·4 or 30·5 m (85, 90 or 100 ft). Typical of the latter was Birmingham's No. 43, shown in our first illustration, with built-in hydraulic levelling jacks. The other illustration shows a special 45·7 m (150 ft) appliance with manual jacks delivered to Kingston-upon-Hull, also in 1935.

F6 Pump Escape
Scammell (GB) 1935

Two years after the Borough of Watford's first Scammell appliance was delivered a second order was placed. This was for an improved model, mounted on a special dropframe chassis and designated Type F6. Equipment, however, was almost identical to the previous machine. Mechanical improvements included a 6-speed constant-mesh gearbox with power take-off and lubrication by means of a submerged gear-type pump, and a double-reduction rear axle. Dewandre compressed air brakes were fitted. As it was delivered in 1935, it was appropriately named *Silver Jubilee* to commemorate the 25-year reign of George V and Mary. It was passed to the National Fire Service in 1941 and to the newly formed Hertfordshire Fire Brigade in 1948. Two years later it was purchased by Metro-Goldwyn-Mayer Ltd, Elstree, for use as the studio appliance before being sold to a Shepherds Bush breaker in 1954. Later that year it was bought by Annis & Co. Ltd, hauliers, of Hayes, Middlesex, and converted into a heavy haulage tractor. It now lies derelict in North London.

FK6 Motor Pump
Leyland (GB) 1936

Introduced towards the end of 1935, the Leyland FK6 was a 6-cylinder petrol-engined machine with rear-mounted pump. A variation of it was the FK7, which had a pump mounted amidships. Both had a wheelbase of 3·5 m (11½ ft). Power output was 62 bhp, the transmission employed a single-plate clutch, 4-speed box and fully-floating overhead-worm rear axle and hydraulic brakes acted on all wheels. This particular machine had a 2,273 lpm (500 gpm) Gwynne pump and 'Braidwood' body, unusual for an appliance of such late vintage.

LG3000 Airfield Fire Crash Tender
Mercedes-Benz (D) 1936

As part of its preparations for hostilities, the German Government backed a number of new designs and improved fire-fighting methods during the period 1934-8. There were particularly important advances in the field of aircraft fire-fighting, one of the most advanced designs being a fully-enclosed appliance based on a bonneted Mercedes-Benz LG3000 6×4 military chassis built by Daimler-Benz A.G., Stuttgart, in 1936. This had a 4·8 litre (293 cu in) 4-cylinder diesel engine, and carbon dioxide gas, stored under pressure in various cylinders, was used as an extinguishing agent, not, as in the case of earlier chemical appliances, to project a stream of water over a considerable distance, but to suffocate the fire which relied upon the availability of oxygen to continue burning. Fixed hose reels were employed, each connected to its own gas supply, and a crew compartment was located at the rear of the body.

Motor Pump & Hose Car
Matford (F)   1937

S.A. Matford was founded at Poissy, Seine-et-Oise, in 1935 following a Ford Motor Co. takeover of the former S.A. Mathis factories. The first new model was a 3,048 kg (3 ton) bonneted truck using mainly Ford components and in 1937 a combination motor pump and hose car with Drouville body and equipment was supplied to the Gourgoin Brigade. Although of open design, this had a double front crew compartment and carried a 1,500 lpm (330 gpm) pump and twin demountable hose reels at the rear. Matfords appear to have used Ford components of various vintages, this particular appliance carrying a 1935 pattern bonnet and 1937 style grille.

'Light Six' Pump Escape
Dennis (GB)   1938

Also known as the 'Big Ace', the Dennis 'Light Six' was announced in 1936, having a 6-cylinder petrol engine of 37·2 hp, single-plate clutch, 4-speed gearbox and spiral-bevel rear axle. It remained in production until 1939. This particular example, delivered to the County Borough of Reading Fire Brigade in 1938, having a 2,270/3,180 lpm (500/700 gpm) 2-stage turbine pump at the rear. Until 1947 it was stationed at the brigade's Caversham Road station but was then transferred to the new Berkshire & Reading Brigade and stationed at various points throughout the county before leaving the service in the early 1960s when it was acquired by Sir Nicholas Williamson, of Mortimer, near Reading, and presented to the Berkshire Veteran Fire Engine Society who now maintain it.

Pumper & Searchlight Truck
Ward LaFrance (USA) 1938

The Fire Apparatus Division of the Ward LaFrance Truck Corp., Elmira, New York State, introduced a new bonneted range of custom appliances in 1937, this combination pumper & searchlight truck being delivered the following year. At this time, many American designs followed the same particular pattern, with open cabs, mid-mounted pumps and standee-type crew area at the rear. Fire department liveries were also far more flamboyant than their European counterparts and certainly more colourful, this particular appliance being finished in yellow.

Model 'V-C' Pumper
Ahrens-Fox (USA) 1939

The Ahrens-Fox Fire Engine Co. went into liquidation in 1936 but was rescued, in part anyway, by the LeBlond-Schacht Truck Co., also of Cincinnati, Ohio, who continued to build Ahrens-Fox appliances to special order. One of these, delivered in 1939, was a 2,270 lpm (500 gpm) Model 'V-C' enclosed pumper incorporating the very latest in streamlined thinking and including a cab along passenger car lines. In common with other American marques, the pump was mounted amidships and a standee crew area provided at the rear. The Ahrens-Fox operation continued along these lines until 1951 when it was purchased by one Walter Walkenhorst, a local businessman, who attempted, largely unsuccessfully, to streamline the operation before re-selling the following year.

'Stork' Utility Trailer Pump
Beresford (GB)   1939

1939 was to see renewed hostilities in Europe and an increased requirement for utility appliances of all types. Trailer pumps were perhaps the most ubiquitous being capable of use with any type of vehicle and one of the most memorable of this type of appliance was the 545 lpm (120 gpm) Beresford 'Stork' powered by an industrial version of the popular 747 cc (45½ cu in) Austin 'Seven' 4-cylinder petrol engine. When required, the pump itself could be lifted clear of its 2-wheeled carriage so that it could remain in use while a second pump was collected or maintenance could be undertaken on one pump without losing a vital trailer unit. This particular 'Stork' was used for many years by Joseph Lucas Ltd, Birmingham, West Midlands, before becoming part of the West Midlands Fire Service Collection.

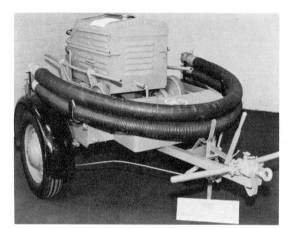

'Light Four' Pump Escape
Dennis (GB) 1939

The attractively styled 'Light Four' appliance appeared in 1936 and remained in production until the end of World War II. This 1939 model with fully-enclosed cab was supplied to the Borough of Watford Fire Brigade, having a 2-stage Dennis No. 2 turbine pump, producing 2,270/2,730 lpm (500/600 gpm), at the rear. Power was provided by a Dennis 'C'-Type side-valve petrol engine producing some 24·8 hp while transmission was effected via a single dry-plate clutch, 4- or 5-speed gearbox and spiral-bevel rear axle. Wheelbase was 3·3 m (131 in).

V8 Utility Car
Ford (GB)   1939

As frantic efforts were made to locate suitable vehicles for fire protection work, the British Home Office even requisitioned a number of new Ford V8 station wagons, painting them utility grey, fitting a ladder rack and rudimentary fire-fighting equipment such as extinguishers, hand pumps and buckets. These were 'first line' appliances, often used in conjunction with a 2-wheeled trailer pump to ascertain the importance of any conflagration before committing more valuable equipment.

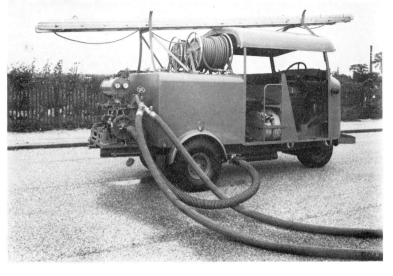

MH.6 Utility Pump Water Tender
Scammell (GB) 1939

Other appliances were constructed on even more unlikely chassis. The 6,096 kg (6 ton) 3-wheeled Scammel MH.6 mechanical horse was one of these, featuring a double open cab and 1,590-litre (350-gallon) water tank. It could turn almost within its own length and had an extended chassis frame to take the transverse 227 lpm (50 gpm) power take-off (pto) driven pump. As well as its twin hose reels, the vehicle's main pump could take water either from its own tank as shown or from a nearby hydrant, supplying two main hoses simultaneously. Variations of this appliance were supplied to some industrial brigades after World War II and this particular machine is now said to be awaiting restoration in the hands of a former Scammell employee.

7V Utility Pump Escape
Fordson Thames (GB) 1940

Large numbers of V8 petrol-engined Fordson Thames 7V utility models were supplied by the Ford Motor Co. Ltd, Dagenham, Essex, to Home Office requirements early in World War II. The earliest of these acted merely as escape carriers in conjunction with 2-wheeled trailer pumps which necessitated the fitting of a specially extended coupling in order to clear the escape wheels. This, however, necessitated the removal of the trailer pump before the escape ladder could be used so many of these tenders were subsequently equipped with a front-mounted pto-driven pump as shown. Based at the Middlesex Fire Brigade's old Park Royal fire station, this 2,032 kg (2 ton) 3 m (118 in) wheelbase model was used for some years after the war as a 'front line' appliance.

FKT1 Multi-purpose Appliance
Leyland (GB) 1940

For the protection of various military installations, the Royal Army Service Corps took delivery of a number of specially constructed Leyland 'FKT' appliances in 1939 and 1940. This, an FKT1 with Gwynne 2-stage rear-mounted 2,273 lpm (500 gpm) turbine pump, was originally a straight pump escape but was later converted into a multi-purpose machine, with numerous additional ladders and equipment. The 'FKT' range had a 96 bhp 7·7 litre (470 cu in) overhead-valve Leyland petrol engine, single-plate clutch, 4-speed transmission and fully-floating overhead-worm rear axle. Most of these military derivatives had open bodies, double cabs and were delivered in full camouflage paint.

'500'-Series Articulated Aerial Ladder Truck
American LaFrance (USA)   1942

Although much of American industry was now concentrating on the war effort, a limited number of new appliances were built for the protection of major cities. Two such machines were 30·5 m (100 ft) articulated aerial ladder trucks supplied to the Chicago Fire Department. Built by the American LaFrance & Foamite Corp., Elmira, New York State, which had been formed in 1927 following the merger of the American LaFrance Fire Engine Co. and the Foamite-Childs Corp., these were based on tractor derivatives of the impressive '500'-Series appliance which had made its debut in June 1938. This could be had with open or closed 3-man cab, as rigid appliance or artic tractor, using the company's own V12 petrol engine. Unlike other makes of aerial ladder, this was a 4-section light-alloy unit with fixed rear steersman's position which did not have to be removed or detached in order to raise the main ladder. By 1953 the economics of post-war operation had forced the Chicago Fire Department to fit Cummins diesels in both machines.

Turntable Ladder
Dennis (GB) 1942

Fire appliance specialists Dennis Bros Ltd and Merryweather & Sons Ltd teamed up in 1938 to manufacture a 30·5 m (100 ft) turntable ladder to take the place of German-developed Metz designs during hostilities. Using a 9·5 litre (580 cu in) 6-cylinder Meadows 6EX/A petrol engine, single-plate clutch, 4-speed 'crash' box and worm-drive rear axle, this was based on a 4·8 m (189½ in) bonneted chassis with open cab and 4-wheel hydraulic brakes. Main advantage of the Merryweather ladder was that the operator travelled round with the base when it was rotated, ensuring that he was always in a convenient position for handling the controls. This particular example was one of the last to be built for the newly formed National Fire Service.

T214 Forest Fire Appliance
Dodge (USA) 1942

Delivered new to the US Armed Forces in 1942 as a 4-stretcher ambulance, this 762 kg (¾ ton) 4×4 Dodge T214-WC54 was later converted into a forest fire appliance for the use of a small American country brigade. Built originally by the Dodge Division of the Chrysler Motors Corp., Detroit, Michigan, the Wayne all-steel integral ambulance body was cut down to the top of the windshield, the doors were removed and a new low body fitted. This had a central gangway for crew and special spray units for coping with grass fires. Wheelbase was 3·1 m (121 in), the engine was a 6-cylinder petrol unit and a 4-speed transmission was employed.

K4 Utility Turntable Ladder
Austin (GB) 1943

Most famous of all British utility appliances were those mounted on the Austin K2 and K4 chassis/cabs. Built in the Austin Motor Co. Ltd's Longbridge Works, Birmingham, from 1939, the former was a 2,030/3,050 kg (2/3 ton) model supplied in large quantities as an auxiliary towing vehicle while the latter, a 5,080 kg (5 ton) machine, could be found as a motor pump or turntable ladder. Featuring a 3-stage manually-operated 18 m (60 ft) Merryweather ladder, the turntable version had a covered crew area behind the cab and was usually fitted with a pto-driven front-mounted pump. Screw-down stabilizer jacks were provided in front of and behind the rear wheels. The K4 had a wheelbase of 2·8 m (111 in), a 3·5 litre (214 cu in) 6-cylinder Austin petrol engine, single dry-plate clutch, 4-speed gearbox, fully-floating spiral bevel rear axle and 4-wheel Lockheed hydraulic brakes.

WOT1 Airfield Fire Crash Tender
Fordson Thames (GB) 1944

The 6×4 3,048 kg (3 ton) Fordson Thames WOT1 was developed specifically for military use and some were delivered to the Royal Air Force as airfield fire crash tenders. These had open soft-top cabs, 2-compartment tanks containing 1,364 litres (300 gallons) of water and 455 litres (100 gallons) of foam compound, two 85 bhp V8 petrol engines (one for propelling the vehicle, the other for pumping), two fixed rear-facing monitors and a folding monitor tower. Wheelbase was 4·2 m (164½ in) and a 4-speed gearbox fitted.

'Comet' 75 Pump Escape
Leyland (GB)   1949

In the late 1940s the British motor industry exported more commercial vehicles than it sold on the home market, happily adapting these to suit overseas conditions. Bodied and equipped by Merryweather & Sons Ltd, this petrol-engined version of the Leyland 'Comet' 75 was delivered to Bombay, India, in 1949, having a very dated 'Braidwood'-style body, mid-mounted pump and open crewcab. Designated Type CP1, the chassis was powered by a 5 litre (305 cu in) 6-cylinder Leyland engine, driving the bevel rear axle via a single dry-plate clutch and 5-speed gearbox. Hydraulic servo-assisted 4-wheel brakes were fitted.

QX Pump Water Tender
Commer (GB)   1950

Shown at the Rootes Group Municipal Exhibition, St Johns Wood, London, in 1950, this 6-cylinder underfloor petrol-engined Commer QX was bodied and equipped by Alfred Miles Ltd, Cheltenham, for the Gloucestershire Fire Service. Announced at the 1948 London Commercial Motor Show, the QX chassis, assembled by Commer Cars Ltd at its Luton factory, quickly gained popularity as a fire appliance. Examples built at the company's new Dunstable truck plant after 1954 often used the famous 3-cylinder 2-stroke diesel engine developed by associate company Tilling-Stevens Ltd, of Maidstone, Kent. The Gloucestershire appliance carried a separately mounted body and crewcab, the latter having jack-knife access doors on each side. A large water tank was mounted amidships with 2,273 lpm (500 gpm) turbine pump at the rear. The engine had a capacity of 4·75 litres (290 cu in) and transmission comprised a single dry-plate clutch, 4-speed gearbox and bevel rear axle.

'700'-Series Pumper
American LaFrance (USA) 1952

American LaFrance's Canadian subsidiary was La-France Fire Engine & Foamite Ltd, Toronto, Ontario, who imported American LaFrance chassis from the United States before bodying and equipping them for the Canadian market. The revolutionary 'cab-forward' '700'-Series, announced by the parent company in December 1945, soon caught on, this example being delivered to Burnaby, British Columbia, in 1952. The vehicle was powered by the company's own V12 petrol engine and the mid-mounted pump had a capacity of some 6,819 lpm (1,500 gpm). The company was so convinced that this design, using a cab with three men in front and two behind, each side of the engine, would succeed that it abandoned all other models. Its hunch was correct and within a few years the '700'-Series formed the backbone of the US fire service.

'S'-Type Dual-Purpose Appliance
Bedford (GB)   1952

With a wheelbase of 4 m (156 in) and a 4·9 litre (300 cu in) 6-cylinder petrol engine, this Bedford 'S'-Type Model SLC3 was fitted out by Alfred Miles Ltd as a dual-purpose pump escape and water tender to the order of the Central Scotland Fire Brigade. Dual-purpose machines are specially suited to rural or small town operation where there may be no mains water supply in some areas. Like the Gloucestershire Commer, crewcab and body were mounted separately, equipment including a 15·25 m (50 ft), demountable Bayley's wheeled escape and 6·25 m (20½ ft) 'Ajax' extension ladder. A large capacity water tank was located amidships and pto-driven pump at the rear.

'SB'-Series Pump Escape
Bedford (GB)   1952

Specially constructed to the requirements of Divisional Officer 'Jimmy' Hole, Chief Engineer, Birmingham Fire & Ambulance Service, this was one of 27 similar appliances supplied between 1951 and 1956 on the Bedford 'SB'-Series passenger chassis, suitably shortened by Messrs Baico to provide a wheelbase of just 3·2 m (126 in) and further adapted for brigade use by Prestage Ltd, Birmingham, West Midlands. The standard 4·9 litre (299 cu in) 6-cylinder overhead-valve Bedford petrol engine, providing a nominal output of 36 hp, was retained, driving a hypoid rear axle via a single dry-plate clutch and 4-speed gearbox. Fitted with a 6-man integral crew-cab body by Wilsdon & Co. Ltd, Solihull, Warwickshire, this was the third such appliance to be delivered, having a rear-mounted Dennis No. 2 turbine pump producing some 1,590/2,270 lpm (350/500 gpm), a 15·25 m (50 ft) John Morris demountable wheeled escape ladder and two shorter 'Ajax' extension ladders.

F12 Pump Water Tender
Dennis (GB)   1954

One of the most familiar specialist appliances delivered to British brigades in early post-war years was
the Dennis F12, a 3·8 m (150 in) wheelbase
forward-control machine powered by a Rolls-Royce
B.80 Mk X straight-8 petrol engine. This made its
debut in 1949 as the slightly longer wheelbase F7
and was available as a pump escape, pump water
tender or dual-purpose machine. Acceleration times
were spectacular to say the least, a typical laden
appliance taking just 45 seconds to reach 97 kmph
(60 mph). The Middlesex Fire Brigade used these
machines almost exclusively at one time, some (as
shown) for use as PWTs, others as PEs. Variations
included the narrow-bodied F8 and turntable ladder
F14.

Series I Industrial Appliance
Land-Rover (GB) 1955

The all-wheel drive Land-Rover, announced by the Rover Co. Ltd, Solihull, Warwickshire, in 1948, was soon found all over the world, undertaking virtually every imaginable task. Some were kitted out as fire appliances for use either in rural areas, where the vehicle's 4-wheel drive came into its own, or on industrial sites where its compactness, particularly in short-wheelbase form, made it highly manoeuvrable. First registered in 1955, this smart little Series I soft-top was used by EMI Ltd, Hayes, Middlesex, for many years. As well as the short extension ladder, it carried a fixed hose reel, pto-driven light-capacity pump, 'first-aid' tank and chemical extinguishers.

'R'-Type Pump Water Tender
Bedford (GB) 1957

Following the demise of the National Fire Service in 1948 and its segregation into mainly county brigades, the Home Office set up the Auxiliary Fire Service to maintain a pool of emergency appliances and highly trained men in the event of a national disaster. Huge numbers of 4×2 Bedford 'S'- and 4×4 Bedford 'R'-Types were ordered with fully-enclosed pump water tender bodies while back-up facilities were provided by military-style trucks of Bedford or Commer manufacture. For many years weekly training sessions such as this were organized for AFS volunteers but eventually the group was disbanded, although many of the appliances were placed in storage for future emergency use. Known as 'Green Goddesses', these re-appeared on British streets during the fireman's strike of the mid-1970s.

'Marquis' Turntable Ladder
Merryweather (GB) 1958

Using a low-loading 'Regent' passenger chassis or specially adapted version of the 'Mercury' goods model, both assembled at the Castle Maudslay, Alcester, Warwickshire, factory of AEC subsidiary The Maudslay Motor Co. Ltd, Merryweathers developed its post-war line of fire appliances. Turntable ladders were based on the 'Regent', in this instance the 'Regent V', to provide the lowest possible centre of gravity, the ladder being a 4-section hydraulic design of light-alloy construction which could be extended to its full height of 30·5 m (100 ft) in just 30 seconds. Compared with the 1942 Dennis shown earlier, the operator now sat at a special console mounted on the nearside of the rotating base and the vehicle carried its own pump, powered by a second diesel engine, both of which were located transversely behind the crewcab. This view of a London Fire Brigade machine also illustrates how the use of unpainted alloy panelling was gaining a hold by the early 1960s.

'Salamander' Airfield Fire Crash Tender
Alvis (GB) 1959

One of the more unusual specialist appliances of the period was the 6×6 Alvis 'Salamander', a derivative of the 'Saracen' armoured car and 'Stalwart' amphibian built by Alvis Ltd, of Coventry, West Midlands. Using six independently-suspended wheels, the first two sets of which were steered, the 'Salamander' was intended solely for operation as an airfield fire crash tender and examples were supplied almost exclusively to military customers. Powered by a rear-mounted Rolls-Royce B.81 6-cylinder petrol engine producing 240 bhp, this particular machine carried a Pyrene Mk VI fully integral light-alloy body containing water and foam compound tanks and fitted with a roof-mounted foam monitor by the Pyrene Co. Ltd, Feltham, Middlesex. Although built for the Royal Ceylon Air Force, it was placed on operational standby at the 1959 Farnborough Air Show prior to shipment.

Model 506.12 Hydraulic Platform
Lancia (I)   1960

Using a 4-wheel drive Model 506.12 chassis constructed by Fabb Automobili Lancia & Cie, Turin, the Milan-based fire equipment manufacturer Bergomi SpA constructed this unusual hydraulic platform unit for aircraft rescue work or water tower use in 1960. Designed to carry four crewmen, this also incorporated a light pump powered by a separate Fiat AR55 petrol engine, a generator, two hose reels and a 400-litre (88-gallon) water tank. The platform could be elevated to a height of 14 m (46 ft) and the chassis had a 195 bhp 6-cylinder petrol engine, 4-speed main gearbox and 2-speed transfer, lockable differentials and full air braking. Maximum speed was said to be 90 kmph (56 mph).

'Firemaster' TFM.2 Turntable Ladder
Leyland (GB)   1960

Although production of the Leyland 'Firemaster' spanned five years from 1958 until 1963, only ten examples were actually built. The 'Firemaster' concept centred upon a flat-topped uncluttered chassis frame rather like a coach chassis, with a horizontal version of the company's 9·8 litre (598 cu in) 150 bhp 0.600 diesel engine located amidships. Transmission was effected via a fluid coupling and Leyland 4-speed semi-automatic epicyclic gearbox with 2-pedal control. A pto-driven 1,590/2,270 lpm (350/500 gpm) Coventry Climax pump was located at the front, between the chassis frame sidemembers, and the example shown here was one of only two 'Firemasters' equipped as turntable ladders. Based on the TFM.2 chassis derivative, bodywork was by Wilsdon & Co. Ltd, Solihull, the 30·5 m (100 ft) hydraulic Magirus ladder being licence-built by David Haydon Ltd, Birmingham. A single-speed rear axle was fitted. Regrettably, the ladder equipment did not come up to expectations and Carmichael & Sons (Worcester) Ltd was brought in to replace it.

C95FD Pumper
Mack (USA) 1960

By 1960 all American fire appliance manufacturers were offering a cab-forward design following American LaFrance's lead back in 1945. In the case of Mack Trucks Inc., Allentown, Pennsylvania, who were still offering their bonneted 'B'-Series, this had been inherited from the C. D. Beck Co., of Sidney, Ohio, which the company had acquired in 1956,

although even Beck themselves had inherited it from the Ahrens-Fox Fire Engine Co., Cincinnati, Ohio, which they had bought out in 1953! Certain fire departments favoured the open-cabbed style such as on this Hale-equipped 4,546 lpm (1,000 gpm) pumper powered by the optional 6-cylinder Mack 'Thermodyne' diesel engine. Note that the old-style curved rear wings had now been replaced by a heavily boxed design incorporating additional storage lockers.

'Nubian' Airfield Fire Crash Tender
Thornycroft (GB) 1960

One of the most prolific builders of airfield fire crash tender chassis during the 1950s and 1960s was Transport Equipment (Thornycroft) Ltd, formerly John I. Thornycroft & Co. Ltd, Basingstoke, Hampshire. The forward-control 'Nubian' was available in 4×4 and 6×6 form, the latter being tremendously popular for fire-fighting work. Supplied to the Ministry of Aviation in 1960, this Model TFA/B80, powered by a 140 bhp 5·7 litre (346 cu in) Rolls-Royce B.80 straight-8 petrol engine, was fitted out by the American Chemetron Corp. as a carbon dioxide/foam appliance, capable of applying a suffocating and cooling layer of foam or CO_2 powder to the flames through a swinging overhead boom and two handlines. While this was largely sufficient for serious aircraft fires twenty years ago, it was nowhere near as fast or as efficient as the latest types.

'TK'-Series Emergency Tender
Bedford (GB) 1961

Built to specifications laid down by Bristol Fire Brigade's Chief Officer, this emergency tender, based on a Bedford 'TK'-Series chassis/scuttle, was designed to cope with all emergencies, even those that might be totally unconnected with fire. Bodied and equipped by HCB Engineering Ltd, Totton, Southampton, Hampshire, this carried various types of lifting and spreading gear, including a special railing spreader for releasing children trapped between railings and equipment for lifting heavy items in confined spaces. Other features included BA (breathing apparatus) gear, floodlights, power tools and both light- and heavy-weight portable oxy-acetylene cutting tools.

'S'-Type Pump Water Tender
Bedford (GB) 1961

Aimed particularly at Third World countries, this 7,112 kg (7 ton) long-wheelbase Bedford 'S'-Type chassis/cab was fitted with a very basic 'Firefly' pump water tender body by Fire Armour Ltd, Willesden, London. All non-essential refinements were omitted, bodywork being steel-panelled with all platforms and bulkheads finished in aluminium chequer plate. A horizontal pattern 'first-aid' hose reel was mounted behind the cylindrical mild-steel water tank and a twin-manifold pto-driven pump fitted at the rear.

'A'-Type Fire Crash Tender
Commer (GB) 1961

Bodied and fully equipped by fire specialists Messrs Kronenburg, of Hedel, Holland, this 7,112 kg (7 ton) petrol-engined Commer 'A'-Type fire crash tender was built to the special order of the Municipality of Maastricht, Holland. The 'A'-Type was developed specifically for fire-fighting work and the integral light-alloy body on this machine incorporated a double crew compartment, 2,000-litre (438-gallon) water tank, 200-litre (44-gallon) foam compound tank and a roof-mounted foam monitor with a foam discharge capacity of 1,000 lpm (220 gpm). Water discharge capacity was 3,000 lpm (657 gpm) at 80 m (262 ft) height. Triple warning sirens were mounted above the front bumper.

Articulated Aerial Ladder Truck
Seagrave (USA) 1961

Although the Seagrave Corp., Columbus, Ohio, offered 30·5 m (100 ft) 4-section aerial ladders with a fixed rear steersman's position from 1961, the old 3-section design with tilting steersman's seat to facilitate ladder erection was still available. This articulated model, with 'conventional' tractor, was delivered to Washington, D.C., that same year, incorporating a wide white band round its middle to make the red-painted appliance more visible at night. Subsequently, this machine also received a white-painted cab roof and a roof over the steersman's position as well as detail modifications to the front end.

'Nubian' Airfield Fire Crash Tender
Thornycroft (GB)   1962

One of a considerable fleet of 6×6 Thornycroft 'Nubian' airfield fire crash tenders with 197 bhp Rolls-Royce B.81 straight-8 high-speed petrol engines supplied to the Australian Department of Civil Aviation was this machine bodied and equipped by Wormald Bros (Victoria) Pty Ltd, one of the largest manufacturers of airfield fire crash tenders in Australia. The roof-mounted foam monitor had a capacity of 18,184 lpm (4,000 gpm) and was capable of projecting this nearly 61 m (200 ft) from the vehicle. Maximum speed was 96·5 kmph (60 mph) and operational weight 13,210 kg (13 tons). Other features included a front 'roo bar, heavy-duty sump guard and short roof-mounted extension ladder.

'Delta' Hydraulic Platform
Dennis (GB) 1963

While the majority of British brigades opted for the somewhat limited turntable ladder for high-level rescue and extinguishing work, American fire departments had adopted the hydraulic platform for such work. One of the first British cities to see this type of appliance in action was Liverpool, which took delivery of a Simon 'Snorkel' based on an unusual low-loading Dennis 'Delta' chassis in 1963. Unlike the TL, the 'Snorkel' could be placed in virtually any position and, if necessary, could even be operated by a crewman travelling in the lifting cage with no attendance below. Built in the UK by Simon Engineering Ltd, Dudley, West Midlands, the appliance shown was an SS/65 model, now replaced by the SS/70, a 21·3 m (70 ft) machine. The 'Delta' chassis incorporated a roomy crewcab and hydraulically-operated extending stabilizers.

'Fire Chief' Pump Escape
Albion-Carmichael (GB)   1966

Developed from the forward-control Albion 'Chieftain' truck chassis, the 'Fire Chief' was a joint production of Albion Motors Ltd, Scotstoun, Glasgow, and Carmichael & Sons (Worcester) Ltd, the earliest examples carrying a normal Leyland Group 'Vista-Vue' cab. Powered by a 125 bhp 6·5 litre (397 cu in) Leyland L.400S 6-cylinder diesel engine coupled to a 6-speed transmission with overdrive, this particular machine was the second of three delivered in 1965, 1966 and 1968 to the Derby County Borough Fire Brigade. With a 6-man integral crewcab, 'first-aid' water tank, Gwynne 2,273 lpm (500 gpm) centrifugal pump and 15·25 m (50 ft) demountable wheeled escape, this was one of only a handful of 'Fire Chiefs' built. Available also in pump water tender and dual-purpose form, it was never popular, the Albion name being more synonymous with goods and passenger types than with premium fire-fighting vehicles.

'15' Mini Pump Escape
Leyland (GB) 1966

In preparation for the opening of new pedestrian precincts in Newcastle-upon Tyne, the Chief Officer of the Newcastle & Gateshead Joint Fire Service required two lightweight mini appliances with turning circles of less than 9·14 m (30 ft), overall body lengths of only 4·1 m (13½ ft), widths of 1·7 m (5½ ft), the capability of negotiating 1 in 14 pedestrian ramps without grounding and an operational weight of less than 3,556 kg (3½ tons). This requirement was answered by two Leyland '15' 762 kg (¾ ton) machines, built at the Coventry factory of Standard-Triumph Ltd, and bodied by Minories Ltd. Each carried a transverse Coventry Climax pump behind the cab, equipment for gaining entry to premises and coupling to nearby hydrants, and a 14 m (46 ft) alloy Lacon ladder. While no provision was made for a 'first-aid' water supply, there was leeway in the design to fit water tanks at a later date, yet still operate within the specified weight limits.

'Nubian Major' Airfield Fire Crash Tender
Thornycroft (GB) 1966

Successor to the 6-wheel drive 'Nubian' was the Thornycroft 'Nubian Major', an altogether heavier and more powerful machine, announced in 1964. Bodied and equipped by Carmichael & Sons (Worcester) Ltd, this 1966 TMA/300 model was typical of the new breed, with separate crewcab and body, almost along the lines of a normal local appliance in the United States. Designed for a gross weight rating of 20,320 kg (20 tons) rather than the 14,224 kg (14 tons) maximum of its predecessor, the

'Nubian Major's' performance was particularly impressive, providing 64 kmph (40 mph) in just 41 seconds. Power came from a 300 bhp Cummins V8-300 diesel engine, which drove all wheels via a fluid clutch, 5-speed semi-automatic SCG epicyclic gearbox and 2-speed transfer box. In the form shown, this appliance was known as the 'Jetranger' 1700, having a water capacity of 6,820 litres (1,500 gallons) and foam capacity of 682 litres (150 gallons). The roof-mounted monitor could deliver 31,820 lpm (7,000 gpm) of foam up to 67 m (220 ft) from the appliance.

84RS Hydraulic Platform
ERF (GB) 1967

Shortly before the 1966 London Commercial Motor Show, ERF Ltd, until then a manufacturer solely of premium heavy goods vehicles, of Sun Works, Sandbach, Cheshire, announced two special fire appliance chassis, one for use as a pump escape or pump water tender, the other as an hydraulic platform. Delivered to the Warwick County Fire Brigade in 1967, this Simon 'Snorkel' unit was typical of the second type, powered by a 235 bhp Rolls-Royce petrol engine and with a 5-speed synchromesh gearbox and midships-mounted power take-off driven from the gearbox via a short shaft. Rear axle drive was disconnected when the pto was in use. Particularly noteworthy was the 'double-ended' crewcab by J. H. Jennings & Son Ltd, an ERF Group company. Soon after this appliance was constructed, a new company—ERF Fire Engineering Ltd—was established at Winsford, Cheshire, to construct complete vehicles to special order.

'Transit' Industrial Motor Pump
Ford (GB) 1967

Based on a 1,524 kg (1½ ton) 3 m (118 in) wheelbase Ford 'Transit' Custom van, this appliance was exhibited at the International Fire Exhibition, Eastbourne, Sussex, in October 1967. Powered by an 85 bhp Ford V4 petrol engine, it was equipped by HCB-Angus Ltd, Totton, Southampton, Hampshire, incorporating a crew area behind the driver with hinged door access on both sides, lift-up tailgate, rear-mounted pto-driven 1,590/1,820 lpm (350/400 gpm) Coventry Climax centrifugal pump, 270/410-litre (60/90-gallon) fibreglass 'first-aid' tank and 8 m (26 ft) 2-section roof-mounted extension ladder. Non-standard mechanical features included heavy-duty battery, radiator and shock absorbers.

'KM'-Series Hydraulic Platform
Bedford (GB) 1969

Much rarer than the Simon 'Snorkel' was the Henry Sissling 'Orbiter', a similar type of hydraulic platform based here on a Bedford 'KM'-Series chassis/cab for the City of Bradford Fire Service. Fire equipment and light-alloy bodywork were supplied by Carmichael & Sons (Worcester) Ltd. The 'Orbiter' had a maximum working height of 21·9 m (72 ft) with a 12·8 m (42 ft) extension. Four hydraulically-operated stabilizing jacks were fitted and the boom itself carried a 76 mm (3 in) pipe supplying water at up to 4,546 lpm (1,000 gpm). A special feature was a water spray curtain fitted around and beneath the platform with control valve operation, thus protecting the operator from excessive heat.

LF1410/52V Airfield Fire Crash Tender
Faun (D) 1970

Operating at a gross weight of 50,800 kg (50 tons), the Faun LF1410/52V rear-engined 8-wheel drive airfield fire crash tender was one of the most powerful machines of its kind when introduced by Faun-Werke GmbH Nuremburg, in 1970. It was powered by a 1,000 bhp Daimler-Benz MB838E V10 diesel engine with ZF torque-converter, 4-speed power-shift box, transfer box and differential lock, providing a 0 to 80 kmph (50 mph) acceleration figure within 38 seconds and 0 to 100 kmph (62 mph) in 65 seconds. Metz fire-fighting equipment comprised an 18,000 litre (3,960-gallon) water tank and 2,000 litres (440 gallons) of foam compound projected by both front- and roof-mounted monitors.

'Ambassador' Pumper & Command Tower
Ward LaFrance (USA) 1970

In 1968 the Ward LaFrance Truck Corp. introduced its unique command tower option for all custom and commercial chassis pumpers. This 4,546 lpm (1,000 gpm) 'Ambassador' pumper delivered to the Wheeling Fire Dept in 1970 was thus fitted, comprising a full-width platform with integral turret pipe elevating to a maximum height of 6·7 m (22 ft) for use as a water, command or floodlight tower. Lifted on vertical hydraulic rams, the command tower access ladder extended automatically with the platform. The angled lower windshield on the 'Ambassador' range provided exceptionally good forward visibility. This model was powered by a 280 bhp 6-cylinder Waukesha F817G petrol engine driving the rear wheels via a twin-plate clutch and Spicer 5-speed synchromesh transmission.

Aerial Ladder
Pirsch (USA) 1971

A manufacturer of fire-fighting equipment for many years and of complete appliances from 1926, the Peter Pirsch & Sons Co., Kenosha, Wisconsin, supplied this 4-stage rear-mounted 30·5 m (100 ft) aerial ladder to the Shorewood, Wisconsin, Fire Dept in 1971. The first rear-mounted aerial ladders had appeared back in 1967 in an attempt to make such appliances more compact and thus more manoeuvrable, previous aerial ladders being front-mounted, resulting in either a very long rear overhang or the use of tractor-trailer combinations. The cutaway rear cab roof and encircling handrail were regular Pirsch features. Extra light-alloy ground ladders were carried within the body as shown, loaded from the rear.

R-300 Pumper & Aerial Ladder
Duplex (USA) 1972

Despite an increasing interest in rear-mounted aerial ladders in the United States, some front-mounted units were still being supplied, a combination pumper and aerial ladder such as this Bean-equipped Duplex R-300 being particularly unusual. Built by the John Bean Division of the FMC Corp., Lansing, Michigan, on a chassis/cab assembled by the Duplex Division of the Warner & Swasey Co., also of Lansing, it was delivered to the City of Speedway Fire Department, Indiana, in August 1972. The aerial ladder was a 25·9 m (85 ft) model, while the midships-mounted pump had a capacity of 5,683 lpm (1,250 gpm). A 2,273-litre (500-gallon) water tank was also carried and the vehicle was powered by an 8V71 2-stroke Detroit Diesel V8 driving the rear wheels through an Allison HT-70 automatic transmission. Although the cab was also supplied by Duplex, this was considerably modified by John Bean who added chrome strips and stainless steel embellishment.

DL30HF Turntable Ladder
Magirus-Deutz (D)   1972

Despite the increasing popularity of the hydraulic platform, a number of which were shown at the International Fire Exhibition, Eastbourne, Sussex, in 1972, Magirus Brandschutz Technik, Ulm, West Germany, a specialist division of Magirus-Deutz A.G., continued to exhibit and demonstrate examples of its world-famous turntable ladder range. Based on a Magirus-Deutz 170D77 forward-control rigid chassis with 6-man crewcab, this 30 m (98 ft) DL30HF unit was destined for the Frankfurt Fire Brigade. As well as its escape ladder facilities, it was also rigged out as a crane, the lifting equipment being positioned just in front of the windshield for travelling. As well as this, a special rescue cage was carried behind the cab and the same lifting equipment used to attach this to hooks at the head of the ladder. A Continental-style demountable hose reel was carried at the rear.

'Vantage' Prototype Pumper
Ward LaFrance (USA)   1972

Some specialist American manufacturers were now looking for something different and were considering more futuristic designs. The 'Vantage', developed jointly as a design exercise by the Ward LaFrance Truck Corp. and the United States Steel Corp., was built only as a prototype, its most striking feature being a 7-man 4-door tiltcab constructed around a welded steel 'safety cage' frame. Known officially as the Series 72, it was based on a Ford C-900 commercial chassis powered by a 5·8 litre (354 cu in) Ford V8 petrol engine. Fire-fighting equipment included a 4,546 lpm (1,000 gpm) 2-stage pump located amidships, pump control panel concealed behind a roll-up door and a 3,409-litre (750-gallon) 'first-aid' tank. Although this appliance caused great interest at that year's IAFC Conference in Cleveland, Ohio, it proceeded no further. The prototype was sold three years later to the Coventry Volunteer Fire Co. Inc., Coventry, New York State.

'Mandator' Foam Tender
AEC (GB) 1974

Using a modified 14,224 kg (14 tons) GVW 3·7 m (145 in) wheelbase AEC 'Mandator' rigid chassis assembled at the Southall, plant of the British Leyland Motor Corp.'s Truck & Bus Division, Chubb Fire Security Ltd, Feltham, supplied this Pyrene FB140/10 foam tender as an initial strike appliance to the requirements of Shell UK Limited's Shell Haven oil terminal in Essex. Retaining its 'Ergomatic' tiltcab and powered by a 220 bhp AEC AV760 6-cylinder diesel engine coupled to a single dry-plate clutch and 6-speed constant-mesh box with overdrive, this machine had a 3-man crew compartment, massive roof-mounted foam monitor with a discharge capacity of 57,280 lpm (12,600 gpm) and provision for four 64 mm (2½ in) handlines. To make up the foam, water was drawn through an 8-way hydrant inlet manifold at the rear of the machine and mixed with foam compound taken from a 3,864-litre (850-gallon) tank. Pump pressure was provided by a rear-mounted pto-driven 6,820 lpm (1,500 gpm) single-stage Godiva centrifugal pump.

'TK'-Series Pump Water Tender
Bedford (GB)   1974

In 1974 the Oxfordshire Fire Service took delivery of three Bedford 'TK'-Series pump water tenders bodied and equipped by HCB-Angus Ltd. Each was based on a shortened 3·6 m (141 in) wheelbase KGS Model fitted with a 'high output' version of the company's 7·64 litre (466 cu in) diesel engine. The timber-framed integral composite bodies each contained a 1,818-litre (400-gallon) 'first-aid' tank and rear-mounted Godiva UMP50 high-stroke low-pressure light-alloy pump discharging via two hose reels. Other equipment included a portable pump, 10·6 m (35 ft) extension ladder, a 4·3 m (14 ft) alloy ladder and a short roof ladder. These machines were intended for 'front line' rural operation and were quickly joined by others of similar layout.

'Pathfinder' Airfield Fire Crash Tender
Chubb (GB) 1974

Winner of a Design Council Engineering Award in 1974, the Chubb 'Pathfinder' is one of the most advanced appliances of its type, incorporating many unique and revolutionary features. This was the first of the 'new generation' airfield fire crash tenders, designed to combat the worst possible emergencies involving 'Jumbo' aircraft. Constructed by Chubb Fire Security Ltd on a rear-engined Boughton 'Griffin' 6×6 chassis assembled by Reynolds Boughton Chassis Ltd, Winkleigh, Devon, this has a central driving position and 5-man crew accommodation, although in an emergency it can be operated by just one man. The massive roof-mounted monitor is capable of projecting a tremendous 61,370 lpm (13,500 gpm) of foam a distance of up to 91 m (300 ft), while two handlines operating simultaneously can discharge 4,546 lpm (1,000 gpm) each. Power comes from a 635 bhp V16 Detroit Diesel. Delivered in 1974, the example shown was the second such appliance supplied to the Massachusetts Port Authority for use at Logan International Airport, Boston.

'D'-Series Pump Water Tender
Dennis (GB) 1974

Purpose-built for use in city or rural areas where its short wheelbase and narrow width made it highly manoeuvrable, the Dennis 'D'-Series pump water tender had a timber-framed aluminium-panelled body containing a 'first-aid' tank of between 455 and 1,818 litres (100 and 400 gallons) capacity. Offered with either a Rolls-Royce B.61 petrol or Perkins T6.354 diesel engine with 5-speed manual or automatic transmission, it was fitted with a 2,273 lpm (500 gpm) rear-mounted No. 2 Dennis pump. Delivered to the London Fire Brigade in 1974, this example had a 6-man crewcab and although suitable brackets were carried had no main extension ladder.

C44 Airfield Fire Crash Tender
Unipower (GB)   1974

Unipower Ltd, Perivale, Middlesex, had had many years experience in the design and construction of specialist off-road vehicles when the 4-wheel drive C44 was introduced in the early 1970s. Developed from the earlier 'Invader', which had been the company's first forward-control model, this was rated at 16,260 or 20,320 kg (16 or 20 tons) GVW and the few examples that were built all formed the basis of airfield fire crash tenders, bodied and equipped by Chubb, Merryweather or HCB-Angus. Delivered to Swansea Municipal Airport, South Wales, in 1974, this was the Merryweather version with compact 4-door crewcab and low-line body. It was powered by a front-mounted 365 bhp turbocharged Cummins NTF365 6-cylinder diesel engine driving both axles via a fully-automatic 5-speed main box, single-speed transfer and air-operated power take-off which enabled the main pump to be actuated or shut off even while the vehicle was in motion. A maximum speed of 104 kmph (65 mph) was claimed.

Mini Pump Water Tender
Volkswagen (D)   1974

Specially developed as a 'front line' appliance, particularly for use in pedestrian areas, this modified German-built Volkswagen double-cab pick-up was bodied and equipped by Branbridge Fire & Security Equipment Ltd, Tunbridge Wells, Kent, in 1974. Designed to accommodate a crew of four, it had a Godiva ACP pump mounted behind the rear-mounted engine driven off a specially designed power take-off. Water capacity was 910 litres (200 gallons), carried in two strategically placed 455-litre (100-gallon) fibreglass tanks. Pump output was 1,590/1,820 lpm (350/400 gpm) via two 64 mm (2½ in) screw-down outlets or one 'first-aid' hose reel located transversely at the rear in true Continental style. Only a couple of these vehicles appear to have been built.

'Pacesetter' Pump Water Tender
Chubb (GB) 1975

Launched at the Interfire Exhibition, Olympia, London, in 1975, the Chubb 'Pacesetter' was designed by Chubb Fire Security Ltd and Loughborough Consultants Ltd, a company set up by Loughborough University of Technology, in collaboration with the Merseyside Fire Brigade. Of similar concept to the earlier Leyland 'Firemaster', but remaining only a design exercise, the 'Pacesetter' called for a 238 bhp rear-mounted 6V71 V6 Detroit Diesel engine, Allison TC470 4-speed torque-converter automatic transmission and front-mounted 4,546 lpm (1,000 gpm) pto-driven 2-stage Godiva UFP centrifugal pump. Mounted on a Boughton 'Scorpio' high-performance chassis, which could accelerate to 64 kmph (40 mph) in just 18 seconds and had a top speed of 121 kmph (75 mph), this vehicle's integral fibreglass and light-alloy body incorporated a 6-man easy-access crew compartment constructed from a single-piece fibreglass moulding. Step height was only 457 mm (18 in) and triple-fold power-operated doors were located on the nearside. Gross weight was 10,160 kg (10 tons).

Airfield Rapid Intervention Vehicle
Range-Rover (GB)   1975

Designed by Gloster Saro Ltd, a Hawker Siddeley company, of Hucclecote, Gloucestershire, in conjunction with the Ministry of Defence for use by the Royal Air Force, this 6×4 conversion of British Leyland's high-performance Range-Rover was constructed as a rapid intervention fire/rescue vehicle. Later offered also on the civilian market, this had a 4-door cab for a 4-man crew, the rear of the integral body containing a 909-litre (200-gallon) water tank connected to a 909 lpm (200 gpm) Godiva Type UFP Mk 6 pump located between the rear seats. Using the standard 3·5 litre (214 cu in) Rover V8 petrol engine, both front and rear axle drive remained permanently engaged while the third axle was of the trailing type. Fully equipped, the gross vehicle weight worked out in the region of 4,064 kg (4 tons).

W300 Airfield Rapid Intervention Vehicle
Dodge (USA) 1977

Using a specially imported American-built Dodge W300 'Power Wagon' with 4-wheel drive, HCB-Angus Ltd developed its 'Firestreak' rapid intervention vehicle for airfield fire and rescue duties in 1974, the example shown here being constructed some three years later. Available with left- or right-hand drive, this had a Chrysler V8 petrol engine developing up to 230 bhp, a 3-speed automatic transmission and a maximum speed in excess of 113 kmph (70 mph), acceleration to 80 kmph (50 mph) being obtained in only 15 seconds. A pre-mixed water/foam solution was carried in a 1,090-litre (240-gallon) fibreglass tank and, as in the case of the Range-Rover, a Godiva Type UFP Mk 6 single-stage centrifugal pump was fitted. The roof-mounted foam monitor was operated from within the cab by an enclosed cable system, having a maximum output of 12,000 lpm (2,640 gpm). Two handlines were also provided.

'L'-Series Aerial Ladder
Oshkosh (USA)   1977

Designed specifically to accommodate the special requirements of aerial ladder, hydraulic platform and water tower manufacturers, the Oshkosh 'L'-Series, with an overall height from ground to top of cab of just 1·8 m (6 ft), was announced in 1977. Con-structed by the Oshkosh Truck Corp., Oshkosh, Wisconsin, the example shown was one of the first to be delivered, having a 30·5 m (100 ft) 4-stage rear-mounted aerial ladder. It was of double-drive 6-wheeled layout, powered by a 350 bhp engine located between the low-line cab and rear-facing crew area. A 4-speed automatic transmission was used.

'M'-Series Airfield Fire Crash Tender
Oshkosh (USA) 1977

That same year the Oshkosh Truck Corp. also introduced its largest and most powerful appliance to date—an 8×8 twin-engined version of its 'M'-Series airfield fire crash tender. This has an overall width of 3 m (10 ft) and approximate height of 4·3 m (14 ft). Water capacity ranges from 18,184 litres (4,000 gallons) up to 27,277 litres (6,000 gallons), while pump capacity is 9,092 lpm (2,000 gpm). The engines, providing a total output of 860 bhp, each drive two axles via a powershift transmission. As well as this version delivered to the Memphis Fire Department, others went to the Tulsa International Airport and the Boeing Co., while the USAF ordered no less than fifty-nine of the biggest units.

Model RIV Airfield Fire Crash Tender
Walter (USA)   1977

The Walter Motor Truck Co., Voorheesville, New York State, offered complete airfield fire crash tenders for some years. This Model RIV 4×4 unit had a 430 bhp rear-mounted diesel engine providing an acceleration time from 0 to 80 kmph (50 mph) of just 18 seconds. Fire-extinguishing equipment included both foam and carbon dioxide agents, the roof-mounted monitor having a capacity of 1,137 lpm (250 gpm) of foam or 6·8 kg per second (15 lb per second) of carbon dioxide powder. In fully equipped form, it weighed in at 8,380 kg (8½ tons). Like other custom-built Walters, this machine incorporated the company's unique '4-Point Positive Drive' system using an automatic lock and torque proportioning differential permitting free differential speed action while dividing the vehicle's torque in proportion to the relative traction of the wheels. Thus, the wheel with the greatest traction receives the most torque and if one has traction none of the others will slip.

D1317 Pump Water Tender
Ford (GB)   1979

This Ford D1317 was specially developed for fire appliance use by the Ford Motor Co. Ltd's Special Vehicle Engineering Department at Langley, Berkshire, complying with all Home Office requirements. The normal gross weight of 16,260 kg (16 tons) has been reduced to 13,210 kg (13 tons) and front and rear axles down-plated to 4,826 and 9,144 kg (4¾ and 9 tons) respectively. Power is provided by a 168 bhp Perkins V8 diesel engine and the integral light-alloy body by Cheshire Fire Engineering Ltd contains a 1,818-litre (400-gallon) 'first-aid' tank and portable pump. A 10·7 m (35 ft) 2-section alloy extension ladder is also carried by this appliance which went to the Shropshire Fire Service in 1979.

'Javelin' Airfield Fire Crash Tender
Gloster Saro (GB)   1979

Resulting from a four year study by the British Airports Authority of its future fire-fighting requirements, the Gloster Saro 'Javelin' is a high-mobility rear-engined 6×6 machine with the ability to shift its 28,450 kg (28 tons) from rest to 80 kmph (50 mph) in less than 40 seconds. Constructed on a centre-drive Boughton 'Taurus' chassis, this appliance has a 14 litre (852 cu in) turbocharged V12 Detroit Diesel engine kicking out some 553 bhp via a 5-speed Boughton automatic transmission. Above the integral 4-man cab is a foam monitor capable of discharging some 45,460 lpm (10,000 gpm) of aerated foam and water. An extra piece of equipment, not fitted on earlier 'Jumbo' appliances, is a rotating aerial ladder and rescue cage, recessed into the roof of the body. Maximum working height for this is 9·8 m (32 ft), enough to reach the centre engines of wide-bodied DC-10 and TriStar aircraft, and no stabilizer jacks are required when this is in operation.

'CX'-Series Airfield Fire Crash Tender
Shelvoke (GB) 1979

During the 1970s former municipal vehicle builder Shelvoke & Drewry Ltd, of Letchworth, developed its SPV range of special purpose vehicles. These were constructed to virtually any specification with a definite accent on fire-fighting machinery. The 'CX'-Series is for airfield fire crash tender applications, available in 4-wheel drive form for 13,720, 17,270 and 22,350 kg (13½, 17 and 22 tons) GVW. These early Merryweather-equipped appliances based on the lightest model were added to the fleet of the Norwich Airport Fire Service in 1979. As in all 'CX'-Series chassis, the 235 bhp B.81SV Rolls-Royce petrol engine is located behind the cab, ensuring that in-cab noise levels are reduced without encountering the type of problems prevalent with a remote rear-mounted unit. An automatic transmission and torque-converter are fitted and the pto enables the foam monitor pump to engage or disengage while the vehicle is in motion. A maximum speed of 105 kmph (65 mph) is claimed with an acceleration time of 0 to 64 kmph (40 mph) in just 29 seconds.

P3000 Emergency Tender
Stonefield (GB)   1979

Based at Goole, South Humberside, this Stonefield P3000 6×4 on/off-highway emergency tender of the Humberside Fire Brigade was designed and equipped specially for dealing with serious road accidents in the brigade's area. Established by the British Government in 1976 to re-vitalize Scottish industry, the short-lived firm of Stonefield Vehicles Ltd, Cumnock, Ayrshire, produced a unique range of 4×4 and 6×4 chassisless vehicles designed by the late Jim McKelvie who had already set up the first Volvo truck and bus plant in the UK, also in Scotland. Powered by a 5·2 litre (317 cu in) V8 petrol engine, this model had a 3-speed automatic transmission with torque-converter and full power take-off arrangement. The integral body carried a multitude of special equipment, comprising mainly electrically-powered cutting and lifting gear. A special extending mast with blue flashing beacon warned other traffic of the vehicle's presence at an incident.

F125 Hydraulic Platform
Dennis (GB) 1980

Because of its very low centre of gravity and low-line crewcab, the Dennis F125 by Hestair Dennis Ltd, Guildford, Surrey, is ideally suited for use either as an hydraulic platform or as a turntable ladder. To enable the front end of the chassis frame to be dropped to accommodate the cab, the 215 bhp Perkins V8.640 diesel engine and Allison MT-643 fully-automatic transmission are set back a little more than in a conventional chassis. Delivered to the Cheshire Fire Brigade in 1980, this appliance has a timber-framed light-alloy body while the cab is mainly steel-framed with a timber-framed roof.

'WY'-Series Turntable Ladder
Shelvoke (GB) 1981

Shelvoke 'WY'-Series chassis, assembled by the Special Purpose Vehicle Division of Shelvoke & Drewry Ltd, are intended for more specialist operations than as a normal 'front line' appliance, this being well catered for by the company's 'WX'-Series. Available in 13,720 or 16,260 kg (13½ or 16 tons) GVW form, this long-wheelbase crew-cabbed version has been fitted with a re-conditioned 3-section Merryweather turntable ladder by G. & T. Fire Control for the Devon Fire Brigade. For operation at the higher gross weight, this appliance even incorporates the TL's original screw-down stabilizing jacks.

'RS'-Series Pump Water Tender
Dennis (GB)   1982

Announced in 1979, the Dennis 'RS'-Series pump water tender represented the first of a new generation of fire appliances from Hestair Dennis Ltd. Conforming to the Government's Type 'B' water tender specification, the 'RS' was developed from the successful 'R'-Series, incorporating a brand new all-steel cab conforming to proposed EEC impact regulations. Offered with a choice of up to four engines, including both Perkins and Rolls-Royce units, it is built entirely at the Guildford, Surrey, factory, incorporating a modular-type body and a choice of Dennis or Godiva pumps up to 4,546 lpm (1,000 gpm) output. In common with other modern appliances, this model is fitted as standard with Lucas-Girling 'Skidcheck' anti-lock braking, ensuring that maximum brake efficiency is maintained, even in the wettest conditions, where emergency braking can sometimes result in a complete lack of control. The appliance shown was supplied to the West Midlands County Fire Service in 1982.

Acknowledgments

The author would like to thank all those without whose help this book would not have been possible. Special thanks go to Olyslager Organisation BV, MB Transport Photos and members of The Commercial Vehicle & Road Transport Club.

For their generous assistance in providing illustrations I would also like to thank the following: H. Austin Clark Jnr; *Commercial Motor*; I. M. Dean; Engineering in Britain Information Services; Firepics; R. Graham; J. Harrington; A. J. Ingram; J. G. Jeudy; B. Jones; A. Krenn; F. C. Lane; Merryweather & Sons Ltd; Science Museum; P. J. Seaword; J. C. Thompson; Transport Picture Library; B. H. Vanderveen.

Source Books

Aircraft
Armoured Fighting Vehicles
Buses
Commercial Vehicles
Dinghies
Helicopters and Vertical Take-Off Aircraft
Hydrofoils and Hovercraft
Industrial Past
Lifeboats
Locomotives
London Transport
Military Support Vehicles
Miniature and Narrow Gauge Railways
Motor Cars
Motorcycles and Sidecars
Naval Aircraft and Aircraft Carriers
Rolls-Royce
Ships
Small Arms
Submarines and Submersibles
Traction Engines
Tractors and Farm Machinery
Trams
Underground Railways
Windmills and Watermills
World War 1 Weapons and Uniforms
World War 2 Weapons and Uniforms

INDEX